S0-BAV-491

# FOSSILS

**Troll Associates**

# FOSSILS

by Louis Sabin

Illustrated by R. Maccabe

**Troll Associates**

*Library of Congress Cataloging in Publication Data*

Sabin, Louis.
    Fossils.

    Summary: Describes fossils, how they were created,
and what they can tell us about the history of life on
earth.
    1. Paleontology—Juvenile literature.  [1. Fossils.
2. Paleontology]  I. Maccabe, Richard, ill.  II. Title.
QE714.5.S33    1984       560        84-2716
ISBN 0-8167-0228-4 (lib. bdg.)
ISBN 0-8167-0229-2 (pbk.)

In ancient times, when people found dinosaur bones, they didn't know they were finding fossils. They came up with many strange explanations for what they had found. Some people said the bones were placed in the rocks by evil spirits to confuse everyone.

Others said that the bones were created by thunderbolts, by seeds carried on the wind, or by some mysterious force that was trying to create living things out of rocks. Still others said that all the bones had been laid in the Earth at the time of the great flood described in the Bible.

Then, about 200 years ago, scientific-minded people began to study dinosaur bones and other fossils. Soon these scientists, called paleontologists, developed theories about what fossils really are.

Fossils are the remains of plants or animals that lived long ago. They may be bones, shells, scales, or teeth. They may be imprints, or impressions of the plant or animal. They may be footprints or tracks left by the animal. They may even be complete plants or animals preserved perfectly in amber or ice.

Most fossils are incomplete plants or animals found in sedimentary rock. Sedimentary rock is rock that has been built up of layers of sand and mud, called sediment. After millions of years, the weight of the sediment turns the mud to rock. Fossils in sedimentary rock may be *casts* or *molds* of prehistoric plants or animals.

A mold fossil is one in which the outside shape and pattern of the once-living thing has been preserved. For example, when a trilobite, a type of prehistoric shellfish, died, it sank to the sea floor. It was buried beneath layer upon layer of sediment.

After millions of years, water seeped
through the hard sediment and dissolved the
trilobite's shell. But the shape of the shell
remained in the surrounding rock. This is
called a mold fossil. A mold fossil may also
show the bone structure of a fish. Or it may
show animal tracks or the feeding burrows
of worms and shellfish or the outline of a leaf
or flower.

Sometimes, a mold fills up with water that

contains minerals. Later, the water dries up, leaving the minerals behind. When this happens over and over again, the mold gradually becomes filled with minerals that take the exact shape and pattern of the original animal. This kind of fossil is called a cast. The cast of a trilobite looks like a solid stone carving of that ancient sea animal. Other cast fossils include fossil bones, coral, and petrified wood.

The fossils found in amber are very different from casts and molds. They are complete specimens of ancient plants or small insects. Very large animals are never fossilized in amber, because amber starts as sticky liquid that oozes out of evergreen trees. This liquid is called resin, or sap.

Millions of years ago, a prehistoric ant might have been climbing a prehistoric pine tree in search of food when it became trapped in the tree's sticky sap. In time, the sap hardened and became the clear material we call amber. Inside the amber, the insect was perfectly preserved. Any fossil encased in amber looks exactly as the animal looked millions of years ago.

Fossils of complete specimens of larger animals are sometimes found in tar or in ice. For example, saber-toothed cats have been found preserved in tar pits in California. And a perfectly preserved specimen of a

wooly mammoth was found encased in ice in the frozen wastes of Siberia. The mammoth —a tusked ancestor of the modern elephant —lived about 20,000 years ago, but it still had grass in its mouth and stomach!

When a fossil such as a dinosaur bone is discovered, it must be carefully removed to prevent damage. If the fossil was buried deep in the ground or surrounded by hard stone or rock, the paleontologist may first excavate by digging, or even by using dynamite.

Closer to the fossil, smaller hand tools are used. Finally, the fossil is protected with a

coating of plastic or shellac. Then it is covered with wet paper, followed by burlap and plaster of Paris. When this protective jacket is dry, the fossil is turned over, and the process is repeated on the other side. Then the dinosaur bone is shipped to a museum, where it will be studied.

When scientists study a fossil, they have a number of ways to decide how old it is. One way is to compare it with similar fossils found in other places. Another way is to compare it with fossils of other species found nearby.

For example, if two trilobite fossils are very similar, the two trilobites probably lived at about the same time, even if one fossil was discovered far away from the other. And if a jellyfish fossil is found with a trilobite fossil, it is probably safe to assume that the jellyfish lived in the same sea at the same time as the trilobite.

Scientists also get clues about the age of a fossil from the material in which it is found. For example, any fossil found in sedimentary rock that formed 200 million years ago must be 200 million years old.

Scientists have special ways to determine just how old different materials are. One way is to measure the amount of certain elements present in the fossil-bearing layers. Although this method cannot pinpoint the exact year in which a trilobite or dinosaur died, it can give the approximate time period in history. Exact dates aren't particularly important when a scientist is dealing with time periods that lasted for millions, or even billions, of years.

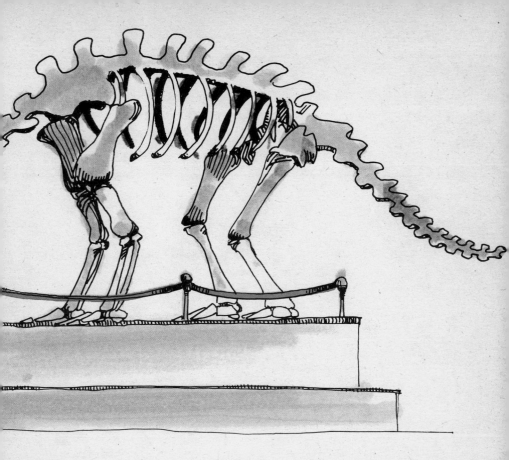

Fossils show us the evolution, or development, of many of today's plants and animals. They also tell us about the many plants and animals that once existed, but which have disappeared from the face of the Earth. The long history of our world is written in rock. By studying fossils, we can learn many things about these changes and the world of long ago.

CANADA

UNITED

Fossils tell us the history of the section of land in which they are found. Suppose the fossil of a sharklike fish is found in the middle of North America. This kind of fish lived only in the oceans and seas. But the middle of North America is nowhere near the sea today. So the fossil shows us that the land surface of North America must have been covered by ocean water during prehistoric times.

STATES

Fossils also tell us about the climate of the past. Suppose the fossil of an ancient fern is found in North America. If ferns of this kind could grow only in tropical climates, scientists would know that the North American climate must have been much warmer in ancient times than it is now.

In the same way, dinosaur fossils found in the Rocky Mountains tell us that the area, which is now cool and dry, was once warm and moist. We know this because dinosaurs needed a year-round supply of water and vegetation. If the climate had not been warm and moist all year, there would not have been enough food for the dinosaurs.

Fossils can also tell us about sudden changes in the ancient climate or land. Sometimes we find large numbers of fossils of young animals with food still in their jaws, or fossils of whole plants with seeds and young leaves and unbroken branches. These young, healthy specimens did not die of old age. They must have been buried by earthquakes, landslides, or volcanic eruptions.

But if the fossils in an area are a mixture of young and old, some of them worn and some broken, it tells us that the specimens died over a period of time. Most likely, there was no sudden climate change and no natural disaster.

Fossils of all kinds have always fascinated people. As far back as 10,000 years ago, people collected them, but we don't know why. We do know that more than 2,000 years ago, the ancient Greeks realized that fossils told the story of the past. When they found rocks containing seashells, they understood that the area had once been covered by the sea. But even though the ancient Greeks were right, their ideas about fossils were not accepted by most people until about 200 years ago.

Today, our museums are filled with fossils of all kinds and all ages. There are fossil molds and fossil casts. There are fossils in amber and fossils found in tar. There are fossil teeth and bones—some of them mounted as complete skeletons to show us the strange and fantastic creatures that roamed the world of long ago. And they tell us what the climate and land were like in prehistoric times.

Each fossil is like a new page in a very special kind of history book. Together, all fossils make up the fossil record—the history of life on the planet Earth, recorded in rock and amber and ice.